Planets and Other Objects in Space

Copyright © by Harcourt, Inc.

All rights reserved. No part of this publication may be reproduced or transmitted in any form or by any means, electronic or mechanical, including photocopy, recording, or any information storage and retrieval system, without permission in writing from the publisher.

Requests for permission to make copies of any part of the work should be addressed to School Permissions and Copyrights, Harcourt, Inc., 6277 Sea Harbor Drive, Orlando, Florida 32887-6777. Fax: 407-345-2418.

HARCOURT and the Harcourt Logo are trademarks of Harcourt, Inc., registered in the United States of America and/or other jurisdictions.

Printed in Mexico

ISBN-13: 978-0-15-362045-4

ISBN-10: 0-15-362045-5

2 3 4 5 6 7 8 9 10 805 16 15 14 13 12 11 10 09 08

Visit *The Learning Site!*
www.harcourtschool.com

Lesson 1

How Do Earth and Its Moon Move?

VOCABULARY
axis
orbit
telescope
moon
phases

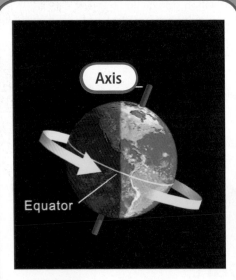

An **axis** is an imaginary line that runs through both poles of Earth.

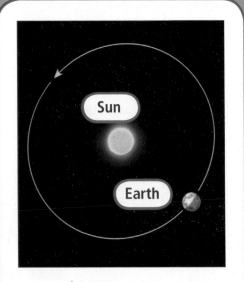

Earth's **orbit** is the path Earth travels around the sun.

A **telescope** helps people see far away objects.

A **moon** is any large body that travels around a planet. Earth has one moon.

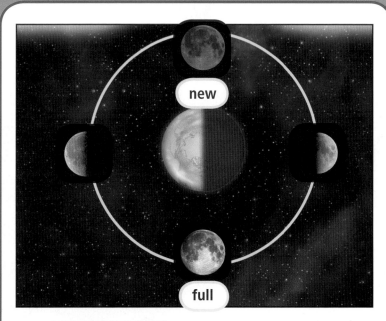

The **moon phases** are the different shapes that the moon appears to have.

READING FOCUS SKILL
SEQUENCE

A sequence is the order in which things happen.
Look for the sequence of events that explains why there are seasons and why the moon appears to change its shape.

Earth's Tilt and the Seasons

Night comes after day. Spring comes after winter. These changes happen because of the two different ways that Earth moves.

Earth turns around on its axis. An **axis** is an imaginary line that goes through Earth. It takes about 24 hours for Earth to make one full turn. During that 24 hours, part of Earth faces the sun. It is day there. The part facing away from the sun has night.

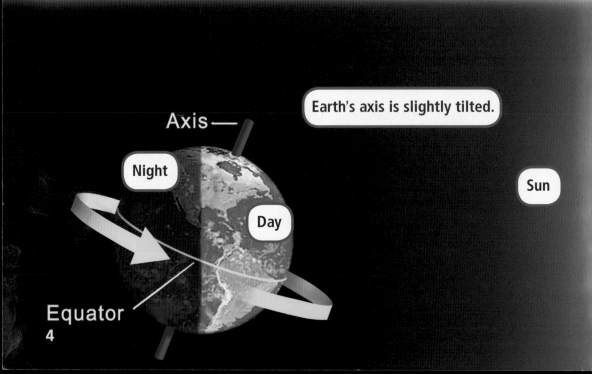

Earth's axis is slightly tilted.

Earth also orbits the sun. An **orbit** is the path of one object in space around another object. As Earth travels, one part tilts toward the sun. This part receives more heat and light. It is summer here. The other part of Earth tilts away from the sun. That part of Earth receives less heat and light. It is winter there.

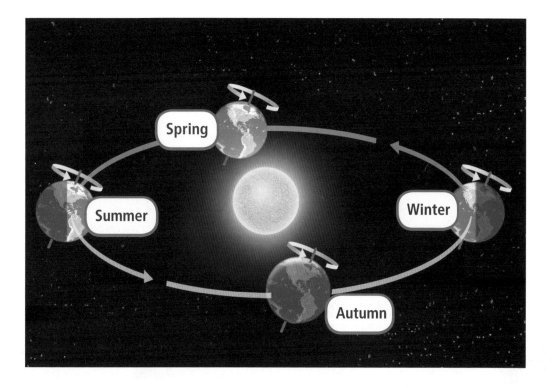

The seasons change as Earth orbits the sun. The part of Earth tilted toward the sun changes.

 Tell what happens next after our part of Earth is titled away from the sun.

Telescopes

The objects in space appear very small to people on Earth. When you look at the night sky, the stars are tiny points of light. How can people study stars if they look so small? Is there a way to make them look bigger?

People use telescopes when they want to look at objects in space. A **telescope** is a device that makes far away things look bigger.

Without telescopes, scientists would not be able to study the planets and other objects in our solar system. Without telescopes, we would not know any details about the objects in our sky.

The first telescopes used lenses to make images larger. Galileo used one of these *refracting* telescopes as early as 1609. Most large, modern telescopes use a mirror to make images larger. Isaac Newton designed and built one of these *reflecting* telescopes in the late 1600s.

 What are the two kinds of telescopes?

Physical Features of the Moon

A **moon** is a large object that orbits a planet. Just like Earth, Earth's moon has physical features. It has a rocky surface, covered in fine dust. It is dry. There are mountains and valleys. There are areas with lots of hills and areas that are flat.

▼ Just like Earth, the moon has mountains.

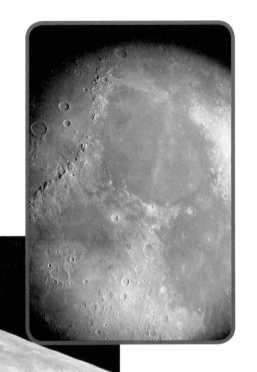

The moon does not have a thick atmosphere, the way Earth does. Because there is no layer of air to protect it, the moon's surface is hit by rocks. These rocks leave craters. A crater is a low area and it has a rim around it.

◀ Dark areas on the moon's surface are large flat areas called *mares*. *Mare* is the Italian word for "sea."

◀ Craters cover the surface of the moon.

Moon Phases

The moon seems to shine. However, the light you see is reflected sunlight.

It takes 29 1/2 days for the moon to complete one orbit. Over 29 1/2 days, the moon's shape seems to change. These different shapes are called **moon phases**. What causes these changes? As the moon orbits, you see different amounts of its surface.

The moon's phases as seen from Earth

New Moon To Full Moon

These phases follow a regular cycle, or pattern. Sometimes you see all of the moon's lit side. This is called a full moon. After that you see less of the moon each day. About 15 days later, you cannot see any of the lit side. This is called a new moon. For about 15 days after the new moon, we see more and more of the moon. Finally, we see all the lit side. Then the cycle begins again.

 What phase of the moon happens after we see less and less of the moon?

Calendars

We use a calendar to measure time. Days, months, and years all come from Earth's movements. Today's calendars have 365 days.

It takes 365 days, 5 hours, 48 minutes, and 45.5 seconds for Earth to orbit the sun. This is called a solar year.

To make up for the extra time in the solar year, we add an extra day to the calendar every four years. We call this year with the extra day a leap year.

◀ A calendar is used to measure time.

Most calendars show days, months, and the year. Some calendars also show moon phases. The phases do not always happen on the same day of every month.

 Tell what occurs every four years on a solar calendar.

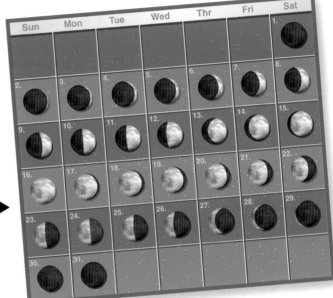

Calendar with phases of the moon ▶

Review

Complete these sequence statements telling how the appearance of the moon changes.

1. The _____ of the moon follow a $29 1/2$-day pattern.

2. During one part of the moon's cycle, you see all the lit side, which is called the _____.

3. Between the full moon and the new moon, we see _____ of the moon each night.

4. After the _____, you see more of the moon each night.

Lesson 2

VOCABULARY
solar system
planet
comet

How Do Objects Move in the Solar System?

A **solar system** is a group of objects in space that travel around a star. The sun is the star in the center of our solar system.

A **planet** is a large object in space that moves around a star. Earth is one of eight planets in our solar system.

A **comet** is a ball of rock, ice, and frozen gas.

READING FOCUS SKILL
COMPARE AND CONTRAST

When you compare and contrast, you tell how things are alike and different.

Look for ways to compare and contrast objects in our solar system.

Our Solar System

A **solar system** is a group of objects in space that travel around a star. The sun is the star in the center of our solar system.

Our solar system has different kinds of objects. There are planets, moons, and asteroids. A **planet** is a large object that moves around a star. A moon is usually smaller. It orbits a planet.

Our solar system has eight planets. They all orbit the sun. Scientists put these planets into two groups. The inner planets are closer to the sun. The outer planets are farther from the sun. These two groups are separated by a ring of small, rocky objects. These objects are asteroids.

 Tell how planets and moons are alike and different.

Our Solar System

Mercury
Venus
Earth
Mars
Jupiter
Saturn
Uranus
Neptune
Sun

Mercury

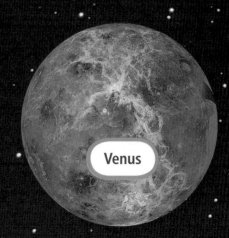

Venus

The Inner Planets

The four inner planets are Mercury, Venus, Earth, and Mars. These planets are alike in many ways. Each has a rocky surface. All are smaller than most of the outer planets. None has more than two moons.

Mercury gets very hot. Mars gets very cold. Only Earth has water on its surface. It is also the only planet that has a lot of oxygen. Water and oxygen let plants and animals live on Earth.

Focus Skill **Tell how the inner planets differ.**

Mars

Earth

The Outer Planets and Pluto

The four outer planets are far from the sun. They are Jupiter, Saturn, Uranus, and Neptune.

Jupiter, Saturn, Uranus, and Neptune are alike in many ways. All these planets are large. They are made mostly of gas. These planets are often called the gas giants. These planets have many moons.

Pluto is different from the gas giants. It is a "dwarf planet." It is small and icy.

 How are the gas giants different from Pluto?

Other Objects in the Solar System

Two other kinds of objects orbit the sun. These objects are asteroids and comets.

Asteroids are made of rock and metal. They are too small to be called planets. Most asteroids orbit between Mars and Jupiter.

Asteroid

A **comet** is a ball of rock, ice, and frozen gas. Most are smaller than asteroids. Comets can change when they come near the sun. Some of the frozen matter turns to gas. This gas looks like a tail.

 Tell one way comets and asteroids are alike.

A comet's fiery tail

Review

 Complete these compare and contrast statements.

1. Moons, planets, and asteroids are all part of the _____.

2. The inner planets are _____ to the sun than the outer planets.

3. The inner planets all have a _____ surface.

4. Asteroids and comets both _____ the sun.

Lesson 3

VOCABULARY
star
sun
constellation
galaxy
universe

What Other Objects Can Be Seen in the Sky?

A **star** is a huge ball of very hot gas.

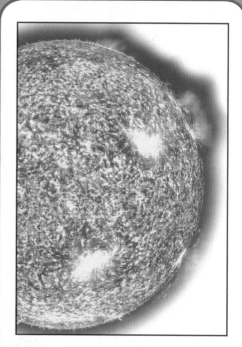

The **sun** is the star in the center of our solar system.

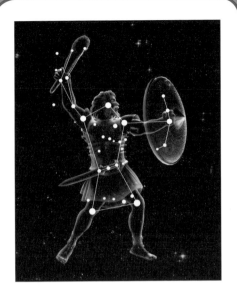

A **constellation** is a group of stars that form an imaginary picture in the sky.

A **galaxy** is a huge system of gases, dust, and stars.

The **universe** is everything in space.

READING FOCUS SKILL
MAIN IDEA AND DETAILS

A **main idea** is what the text is mostly about. **Details** tell more about the **main idea**.

Look for **details** about the sun and other groups of stars.

The Sun and Other Stars

A star is a huge ball of very hot gas. Stars may be different sizes and colors. Red stars are the coolest. Blue stars are the hottest. Yellow stars are between the hottest and the coolest.

Positions of nearby stars

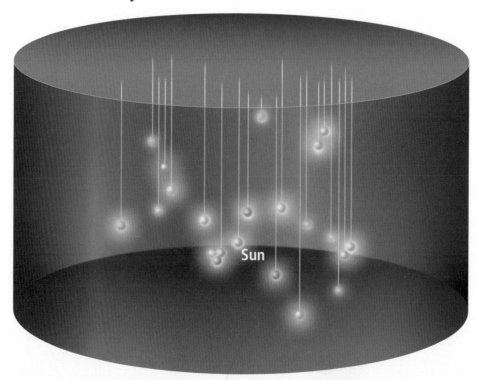

The **sun** is the star at the center of our solar system. It is a medium-sized, yellow star. Most energy on Earth comes from the sun.

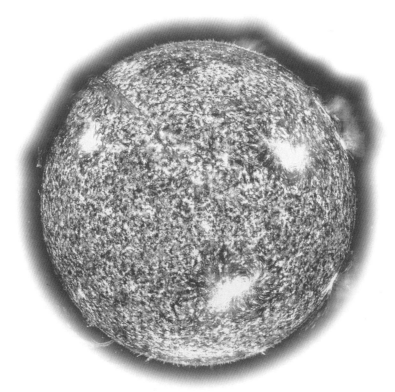

The sun is more than 1 million kilometers (621,000 miles) in diameter.

Look at the red loops in the picture. The loops are solar flares. These are cooler than other parts of the sun.

 Explain what a star's color tells about it.

Groups of Stars

The Big Dipper is part of a constellation. A **constellation** is a group of stars that make an imaginary picture.

A **galaxy** has billions of stars. A galaxy is a huge system of gases, dust, and stars. Our solar system is on the edge of the Milky Way galaxy.

The **universe** is everything in space. It has billions of galaxies.

 What are two ways that people classify groups of stars?

▼ Galaxy

▲ This constellation is called Orion.

Seasonal Star Movements

Each day the sun appears to move in the sky. But the sun does not move. It is Earth that moves through space.

At night the stars seem to move. Their positions appear to change from season to season. But this is due to Earth's movement. As Earth orbits the sun, we see different parts of space in the night-time sky at different times of year.

▲ Constellations in the summer sky

▲ Constellations in the winter sky

 Tell why constellations seem to change their positions in space.

Review

Complete this main idea statement.

1. There are several ways to classify groups of _____.

Complete these detail sentences.

2. The star in the center of our solar system is the _____.

3. Imaginary pictures made of stars are called _____.

4. The _____ is made up of billions of galaxies.

GLOSSARY

axis (AK•sis) The imaginary line that Earth spins around as it rotates

comet (KAHM•it) A ball of rock, ice, and frozen gas in space

constellation (kahn•stuh•LAY•shuhn) A pattern of stars that form an imaginary picture or design in the sky

galaxy (GAL•uhk•see) A huge system of stars, gas, and dust

moon (MOON) Any natural body that revolves around a planet

moon phases (MOON FAYZ•uhz) The different shapes that Earth's moon seems to have

orbit (AWR•bit) The path Earth takes around the sun

planet (PLAN•it) A large object that moves around a star

solar system (SOH•ler SIS•tuhm) A group of objects in space that revolve around a central star

star (STAR) A huge ball of superheated gas

sun (SUHN) The star at the center of our solar system

telescope (TEL•uh•skohp) A device people use to observe distant objects with their eyes

universe (YOO•nuh•vers) Everything that exists in space